SHE WAS LOVED

JOSEFINA DE VASCONCELLOS

Daydreaming
the early years

Bas relief by
Josefina de Vasconcellos
(cast in resin bronze)

She was Loved

Memories of Beatrix Potter

———————

Josefina de Vasconcellos

Titus Wilson

2003

For a wide variety of help, my warmest thanks.

Text copyright ©Josefina de Vasconcellos
Foreword copyright © Mary Burkett
Images copyright © owners
Illustrations © Armitt Trust
Letters from Beatrix Potter copyright © Frederick Warne & Co., 1936, 1938, 1939, 1943. Reproduced by permission of Frederick Warne & Co

ISBN 0 900811 27 7

Designed by Stephen Hebron

Printed in Great Britain by Titus Wilson, Kendal

Dedication

Beloved Lady – Beatrix …
Girl, Woman, old Friend –
I seem to have always known you.
The beauty inherent in your work
has drifted through my life,
like perfume from a bunch
of cottage garden flowers.

Foreword

Mary E. Burkett

MANY BOOKS HAVE been written about the life and works of Beatrix Potter, all with enthusiasm, some the result of much arduous research, but none of the authors, not even her nephew, had ever met her or known her. This account is different. Josefina de Vasconcellos and Beatrix Potter not only knew each other but became great friends. The following pages tell of the frequent visits by Josefina and her husband to Hilltop and Sawrey. They also allow the author to retrace gently her memories (and despite her great age hers is a phenomenal one) and so to reminisce on those meetings seventy years ago, as well as some about herself which haven't as yet been recorded. Ideas trigger visions of events in her own life linking them in similarity. It is as if she is lifting the layers of time and discovering shared experiences, which somehow bound them together.

Although an ardent admirer of Beatrix Potter I was not fortunate enough to have known her. She died in 1943, I came to the Lake District in 1954, but I am lucky enough to be a friend of Josefina's and having known her for over fifty years I can understand the nuances of meaning, aspirations and deeply held opinions that she expresses here. What also helps me write these few words, is having known myself several of the other people who knew Beatrix Potter really well. Two friends in particular gave me vivid impressions of the real Beatrix Potter – and it is the real person that Josefina conjures up. One of those friends was Benita Gaddum, a niece by marriage of William Heelis. She described Beatrix Potter's acerbic sense of humour, lack of pomposity and straightforward manner. How well this last comes over in her letter to both Josefina and Delmar Banner when she is criticising their work – no flannel, no sentimentality, but praise where praise was due: 'Your husband has learnt clouds. Light next please. He has the drawing which is the foundation'.

The second friend May Moore had been a lady's maid to Mrs Hext at Coniston and through her met her friend Mrs Heelis and became a great help to her in rendering personal services such as dressmaking (usually sewing up hems) and even cutting her hair. Mrs Heelis always referred to her as 'the little one' because May was even smaller than she was herself. She always welcomed her and once even gave her an original animal drawing which May, alas, gave to one of her nieces and it was lost. But I must not begin reminiscing. I only want to stress how human the author of the *Tales* was. She was capable of writing to and for children despite the fact she had none of her own, another factor shared with Josefina, and yet Beatrix was not sentimental. Her attitude to the animals she portrayed was meant to be strictly realistic. When I was staging the Beatrix Potter Exhibition at Abbot Hall in the 'sixties the rumour had gone around that she hated children. One day I was in the Gallery and overheard the conversation of two middle-aged women. They were looking at the pictures and speaking in an animated way about their experiences of meeting her when they were growing up in Hawkshead. I asked them about her attitude to children at this point.

'She used to play with us and go hunting for bears in the wood', they told me. 'It was only them toffee-nosed children who were brought from the towns to visit her that she didn't bother with', they added, 'She liked the children and we liked her'.

Of course she could be stern, but she was very sympathetic as can be seen in the many acts of kindness that she was doing for the Banners.

The style of the letters evokes memories of a former age in the way they address each other with an old world formality. In 1937 it was, 'Dear Mrs Banner', in December it developed into 'Dear DH and Pigwig Banner', (Pigwig was her nickname for Josefina). On 17 December 1937, 'My dear Pigwig and Delmar'. It was only in 1938 that, 'My dear Josephina', became a regular form of address.

Both women shared a deep interest in art and of course were practising artists. The helpful remarks the older of the pair gave the younger were gratefully accepted by Josefina. Both women are totally honest in their opinions, their friendship was too deep to allow criticism to spoil their relationship. Mrs Heelis described to Josefina admiringly, 'The amazing show of water colours', by Dick Yeadon. (I remember our showing his work at Abbot Hall some years later.) She deplored his early death at forty. She disapproved of the way art was sometimes taught in schools, teachers stifling the natural originality of the child by superimposing unnecessary influences. Neither woman is afraid of challenging accepted local 'icons'.

'I never thought much of John Peel', was Beatrix Potter's terse comment on hearing the huntsman was once so drunk he nearly fell off his horse. They were individuals and happy to acknowledge their foibles.

'Except for us lunatics who spend money on a view', Mrs Heelis said when discussing the priorities in looking for a house.

Of course we can hardly expect Beatrix Potter to have admired someone who hunted the progeny of Mr Tod, but the attitude was not due to her being in any way superior to local people. All her friends, apart from a few local people, were shepherds or farmhands who can't have shared in any part of her intellectual life, or else they were ardent admirers like the family in Canada with whom she corresponded. Those letters were concerned with growing up, friends and family matters. Josefina was therefore one of a very few people with whom she had any sort of intellectual relationship. She used to quote Shakespeare to her and expected Josefina to know what she was talking about.

Leslie Linder, *The Journal of Beatrix Potter*, his amazing book describing the decoding of her diaries, opened a new window on Beatrix Potter. *She was Loved*, by Josefina de Vasconcellos has opened yet another, in this year of 2003. At 98 years old and frail in body Josefina keeps working, brush or pen poised, modelling clay on her knee. 'I can't turn pages over easily, but I can still hold a chisel or a mallet', she says with a smile. No grumbles, no complaints. 'Am getting used to being able to use my left hand'. In fact her cheerful nature echoes that of Beatrix Potter who when faced with problems used to say 'Make the best of things and keep working'.

Josefina still writes poetry and prose, sculpts, but can no longer dance. But her mind is focussed and dances in original ideas. These two creative twentieth century artists leave posterity a wealth of delights which will go on into the future, to perpetuate their memories and be cherished for ever.

Introduction

STARTED IN 1961 and finished in 2003, this book was written for those who love Beatrix and the magic of her art. Thus, in one flick of a Tail – or raising a Whisker – you become a friend, so I can relate without formality. When I am asked, 'What was she like?' I say – when you meet her, you feel like someone who has always eaten White Bread and suddenly tastes Brown Wholemeal.

I wonder when you first got to know her? Delmar (my life-companion) had never heard of the *Tales*, until I gave them to him as an Engagement Present. It went down well and one of his favourites was *The Tale of Pigling Bland*. The characters of Pigling and Pigwig seemed based on our own. He was cautious, thoughtful and responsible but she danced her way regardless.

A less romantic trait was the fact that when carried away by sudden laughter, a loud snort popped out! From then on I became PIGWIG.

The lonely Child

by Josefina de Vasconcellos
(cast in J B formula)

I

HERE IS A little girl – aware, and wary as any wild creature – scenting the air – listening – hidden under the long table cloth … hearing adult conservation. Whether it was puzzling to, unsuitable for or beyond a small child's understanding, never seems to have troubled her unsympathetic parents. Approach her – as gently as you would a sleeping baby.

Nursery Window at The Boltons by Josefina de Vasconcellos

(mezzo relief in J B formula)

STUDY THE FACES of the Parents of a child, starved of affection, isolated from sensitive understanding. The Father, bristling with deeply fierce sexual frustration, the Mother a tight knot of petty negative conventions – with enough "Accide" to put off a Baboon on heat.

Even in those days of "Children seen but not heard" it was not usual for the daughter of the house to be under the table like a dog – she was not even "seen". Her nursery was a Prison but her Soul flew free.

The Prisoner lets her pet - her only company - fly free
a poetic symbol of body and soul
by Josefina de Vasconcellos
(Cast in J B formula)

$$3$$

DISMISSED FROM STUDY or play with other children, yet her pets gave her companionship and comfort – and no little amusement. From her high lonely window she gained an insight into the sound and movement of wind in the complex of leaves and branches in the fine tall trees of *The Boltons*.

Langdale by Delmar Banner (Private Collection)

———— 4 ————

WITH HER FIRST holiday in the Lake District, a new and compelling passion flushed into her life – and she blossomed, heart and mind resting with calm, containing contours of the fells.

Here was home – a hungry soul was fed.

Josefina de Vasconcellos, Self-portrait (Private Collection)

HER FIRST STORY opened a new door into her life. She met the Warne children and had the fun she had missed. Also – she understood children and knew what they like … the naughtiness of Tom Kitten. How enjoyable is his rude playing with his Mother's best hats.

The naughty daring cheek of Squirrel Nutkin, teasing Old Brown. Peter Rabbit's exciting disobedience. The child in us all wants him to explore. This is where it is not realised that Beatrix herself never lost that innocent naughtiness that delights us in the tales.

She had never lost it and this is one of the reasons why we understood each other in an 'creative' way without words. At much the age when she or her brother were boiling the stinking dead Fox, I was walking along a grassy path with my Mother, tailing behind, observant, and there was a bright brown adder with a V on its head, so quite gently I picked it up. Later on, my mother said 'Why are you walking so funnily?' 'I've got a snake in my pocket and don't want to disturb it' – 'PUT THAT THING DOWN!'

I also was not allowed to have children in to play. Once a friend of my Father's came to the house with his little boy, and left him in the garden. We played very happily and I enjoyed his world but Father never let it happen again.

Watercolour by Beatrix Potter (Private Collection)

MOVE ON FROM naughty pranks to the more serious business of the bad relationship between Tommy Brock and Mr Tod. We enjoy the water trap being set – find Beatrix here too. One day, while we were chatting on local properties for sale, she put on a 'a cloak and dagger dramatic snarling voice' and said 'Old Josna Scriven and I are waiting for each other to die!' The tales were true to life. Her humour was always bubbling up like a spring.

In me she found someone who had hidden lost love and covered her sorrows as she had – and had a secret area of unwritten poems. She was as secretive and reclusive as Gordon Wordsworth – but unlike him, unable to hide away in the woods with the birds. He told me the number of nesting boxes he had placed – but I will not tell, because it will not be believed.

Yes he too saw inside my shell as she did. He took me up into his bedroom – the window was open and in the ivy below the sill was a small deserted nest. 'A Wren' he said, 'but the mice got her eggs'. We stayed sadly quiet – looking at the perfection of the nest and over the garden to the river where Trout sped like shadows and the Dippers quietly bobbed before a dive.

The other dear friends lived on the same magical land nearer The Stepping Stones. Canon Romaine Hervey and Ethel his wife – both the most beautiful and saintly friends I had ever known. Of course, they had a Robin that came in at tea time and flitted between buttered Scones and Maderia cake.

Gordon's Funeral Service was in the barn beamed St Oswald's Church – his grave was near the family stones of the poet. During the service a Robin stayed on a branch of a yew tree above the grave. Nobody seemed to notice it, but we looked each other in the eye, and knew he was there not only for himself – but to represent all the loved birds of a friend.

We all know the Robin in *The Tale of Peter Rabbit* and associate him with our own garden Robin.

When I was seven and had more time in the garden 'my' Robin became very tame – he came into the kitchen when I was there and looked a picture, perched among the white dinner set on the Dresser between a few flights to the table where delicious crumbs were picked up. One day in early June I was on the big lawn surrounded by Apple trees and between them the flights of sweet peas perched on their supports of woven branches.

A chirping and chat of a pair in perplexity issued to request my help. The problem was that their baby was sitting on the lawn and unable to fly. They had tried in vain to teach it. I realised – too well-fed too fat to heave itself off the ground! I took it in my hands – lifted it and launched it for a lob of two feet. It flapped its wings on landing. The parents very pleased – watching from the sweet pea stand. Another try – this time a little higher and a longer launch. It began to co-operate. Flip-flip. By the end of the session I had got him flying the full length of the garden and the delighted parents took him off.

Remember how Beatrix trained the original Peter to jump over a rope?

Robin

Robin arrives –
He says 'It's ME!
My branch, my tree,
I'm trim and free.'

It's YOU, I find –
I'm pleased to see
That you are kind,
Good company –

Your work is play,
So dig day long,
For worms I'll pay
With song –
And steal your heart away.

Watercolour by Beatrix Potter (Private Collection)

THE TALES THEMSELVES, without illustrations, remove you into a world of reality so steeped in Nature that you may smell the earth. I know the characters are authentic animals. Not so with others who strain to copy the originals. The joints are fictitious – pretty caricatures of Mice and Rabbits dressed like toys. Strip off their costumes and underneath find nothing … and the stories tell you more about the Authors than the characters.

Beatrix has studied and drawn the forms of her characters with the anatomical passion of a Naturalist – and the clothes they wear look as though they had worn them for ages. But note – when Mr Tod is surprised we see him escaping in his own natural fur. Their clothes are faded with sun and washing as revealed in the tale of Mrs Tiggywinkle.

If ever you are in Ambleside make the steep ascent of the lane that leads to High Skelghyll (home of Mrs Tiggy's customers), the ancient Farm set cosily in the hillside.

We stayed there for a week during the war. The boys, Brian and Bill (still young enough to be wearing button-up nightshirts) were heard calling to us from their open bedroom window. They were wakened by the strange gobbling noises in the farmyard below.

'What is that Bushy Animal?' It was a large Turkey on the strut among his indifferent hens.

That night Barrow was attacked by the Luftwaffe and we could see the glare of gunfire in the sky above. A bomb (meant for the seaplane factory below in the woods on the shore of Windermere) was dropped near us – luckily in a bog. The whole Farm leapt into the air and down – a memorable night.

Its not a cheerful time. A most peculiar war; for those of us who lived through the last one, it seems different; "bad 1/0 reckon up" as the saying is. And everything in a muddle. There is no use thinking, keep working and make the best of things!

Our own night bombers flew back to base at Barrow, passing over our valley. We listened anxiously to hear if the engines had been damaged, working on a limping beat – would they make the height over the mountain?

One winter evening, we listened, the beat was out of tune – came the flash and explosion of a crash.

The whole valley knew …

The boys bailed out, but the strong wind blew their parachutes against the icy crags and they hung there frozen for 4 days – while shepherds at the greatest peril to their lives, were lowered on ropes to bring down the bodies.

Many others were not discovered until weeks later.

In summer I was near the top of a fell – there in the grass like a beautiful piece of white sculpture lay a man's hand. I hid it under stones lest the ravens should find it. Further on I found a man's leg torn off from the clothing and body. I ran down the fell top speed like a shepherd with my knees bent – leaping becks and stones. Our only phone in the valley at the Tourists Rest got me to the HQ in Barrow. Next day they came to clear the mess.

Whispers in the Grass

Wandering alone
She found a sheltered Plot
with a view
for a Memorial Stone –
She knew, Cumberland men
Killed in the War
would love that spot …

No Stone was raised,
Sheep graze and Shepherds pass,
Yet their dear names are praised
sung by the wind
and whispered in the grass.

Rock Rock a-bye

The shadow of the cradle moved slow on the wall.
The winter night was sleeping with snow soon to fall.
And I my watch was keeping lest Baby wake and cry
With a rock rock a-bye.

There came a sound of thunder with light clear and red
My mind was held in wonder, my heart filled with dread
And like a falling snowflake, A Lad fell from the sky.
With a rock rock a-bye.

He fell upon the hard rock, the high granite wall,
Above the lonely sheep fold and wild waterfall
Alone upon the hard rock he cold and stark will lie
With a rock rock a-bye

O Heaven pray keep mercy on folk such as I
And keep my son from harm lest so young he should die
To love a shepherd's calling and never learn to fly
With a rock rock a-bye.

I wrote music for the above and sang it to a few local friends.

Josefina de Vasconcellos in the forties.

——— 8 ———

AFTER THE WAR Beatrix wandered in the open ground, looking for a site for a memorial to the Cumbria Casualties.

Today, I after years of research have been given a worthy place in the West end of St. Patrick's Church Patterdale for a memorial to the RAF boys whose planes crashed on the fells. Lads from the USA, Australia, New Zealand, Rhodesia, South Africa, Holland, Poland, Czechoslovakia and our own will be remembered in a book. The book will rest upon an Altar from St Martin in the Fields. In the Blitz it was carried down into the Undercroft for Dick Shepphard to give Services in safety. Years later I found it among the dump of stuff locked away in the dark.

It happened when Austen Williams wanted a Sculpture for Refugee Year. We agreed that the Holy Family as Refugees into Egypt would be understood. I said, why not use the Undercroft as a Studio then it can be brought out and exhibited in the Crypt. In spite of its fine proportions and beautiful vaulting and two tomb recesses and the flagged flooring the Undercroft had remained a locked up area where every unwanted thing was dumped.

That it was dark was no problem. *They fled by Night* needed only Moonlight which was provided by a strong lamp on a long flex mounted upon a Lacrosse Pale. I needed a work table and found one cluttered with old books and paint pots which with one shake I swept off onto the floor. A second shake to remove dust revealed a feeling of carved wood under my hands and by the light of my torch – an Altar!

How I got a team of ex-Glider Pilots to help restore the Undercroft and prepare it for a new Exhibition "A Calendar of Life presenting Painting Poetry Sculpture Pottery and Objects of Natural Beauty" at the Vicar's request is another story – and the Altar with a lovely antique wood carving on it took pride of place.

Of course, working in the crypt – as I had done under the time of Austen Williams – always in my working clothes and scrubbing the floor when we ran short of cleaners, I got to know the Regulars because they took me for a char. I had two narrow escapes, one from a drunk vandal and one from a sex offender. The Tarts I got to know when they came into the wash room to smarten up before going on the street. They wanted to find a sympathic listener who would chat on the usual female topics of clothes and diet. When loo paper ran out they wiped their bottoms on the walls. One cannot always be carrying tissues around though in general we all do. Beatrix also would not have turned a cold shoulder. I was known among them as 'a good Tart wasted'. Sometimes a chance comment or turn of words would give a fleeting insight into a soul shut out of a normal life by the cruelty of parents or sheer poverty. Many have hidden treasures that never see the light.

And so it was with Beatrix. A comforting word in childhood would have released the natural poet. Her modesty was too prohibitive. The Nursery Rhyme type jingles in the tales were written in without a thought of the conventional forms and attitudes of poetry. Children's Poets are a dedicated race apart.

—————— 9 ——————

DEATH STOLE HER young love from her … one does not record, one only covers.

In one of our last conversations together she told me how she lost her engagement ring she always wore in the hayfield and the sadness of it – 'I searched and searched'.

I have his umbrella.

Beatrix Potter by Delmar Banner.
This is how she looked when I first met her – at the Eskdale Show.

Whispers in the Grass

MY FIRST MEETING with the tubby old lady with rosy cheeks and the brightest blue eyes – was at the Eskdale Autumn Sheep Show. She was one of the Judges – all breeders themselves – and I joined the silent crowd leaning over the sheep pens in listening and watching concentration, when I could.

But I was very busy (the only 'offcomer' allowed in that whole week) helping the Armstrong family prepare for the Shepherds' Lunch, laid out on long tables in the Shippon next to the Woolpack Inn. Dressed in my usual Sculpting overalls and wearing Clogs, I served huge plates of delicious cold lamb, ham and sliced beetroot, eggs, potato and salad, to the Shepherds seated (with generally two dogs under the benches) in the cool shade.

And there we met, out on the grass already well on its way to be trodden into mud. A farmer friend led me to her at one of the folds for Herdwick Rams (she had already won a silver Cup for one of her Ewes) and introduced. She slowly looked me up and down – observed my Clogs, and was satisfied. Without further ado she said 'Cum to tea on the –th and bring your Painter Husband'. I was extremely thrilled! It seemed too good to be true, and I knew it was the know-how of my farming friend that had done the trick. Almost immediately she was slapped on the back by a Shepherd in congratulation for the Silver Cup 'Well done lass' and nearly fell over. Another farmer told how his father had seen John Peel so drunk he nearly fell off his horse. 'I never thought mooch of John Peel!' was her reply.

It was an important and mostly enjoyable day for so many. The WI Tent was beautifully set out. Needle-work of many kinds – Knitting, Crochet and Lace. Sets of a dozen eggs of matching size and colour presented in rounded Nests of hay. Jars gleaming with colours of jams, preserves, chutney and honey -– home brewed wines of Elder blossom, Elderberry, Parsnip and Burnet etc., hand-made artefacts, pottery and bead-work, their makers waiting for judgement, chatting and having tea, children on the run to see everything and have ice creams. Serious men standing apart, scanning the surrounding fells with field glasses to spot any unlawful interference by types from the towns to the trail, now being walked to a finish, by a man dragging an aniseed rag for the hounds to follow. The night before, some experienced and responsible local men had been out hiding among the rocks with guns (to let off to frighten any possible lurkers to switch the Trail).

Sounds of fast motorbikes came from the west, but of course no suspicion could fall on those who had simply come to make bets. There was some noise and the bookmakers were busy. Owners of hounds with odd or local names were walking their dogs to the starting line. 'The puppy trail' would come later – and then the contest for the best 'shepherd's call'. I entered for that – yes I could take the stand, put my little finger in my left ear hole and throw my voice over the fell. There was no showing off – all quite natural of an old tradition.

Last of all a Walsing Contest! Partners (some a bit unsteady for drink) wearing wellingtons or high boots, danced in the (by now, thick and trodden mud) in the evening light, to the music of Blue Danube by our local Band. The prize was an umbrella. The sheep had been gathered and driven home or loaded into vans.

So ended the last Show of the year.

*

The Moon rose over Bow Fell and all was quiet, save for the trembling call of an Owl in the woods behind the Woolpack Inn.

*

Do you get a feeling that I loved all these people and the place? I think that even visitors from Japan and other far-off places do feel at home here because of the rivers, the steep fell sides, and interesting shapes of trees glowing out from the rocks, so beautifully presented in their art. There is so much I would like to tell, but must keep descriptions only relevant to incidents within my tale.

Well on the appointed day – fine and dry with Damsons already picked and Apples ripe and rosy on the trees – we came to Castle Cottage. The front door faced the back garden and orchard, and we came through two other gardens to find it. It was painted a faded blue, with no letterbox or knocker – so we used our knuckles. A long silence ensued but eventually the sound of Clogs slowly

Josefina and Delmar outside The Bield on their Golden Wedding Anniversary

toddling over flagstones came near – and then stopped behind the door. We kept quite still – it was like waiting outside Tittlemouse's dwelling and hoping she would appear …

This, and some other incidents, highly revealing of a very unusual personality, given to me I have recorded in other books on Beatrix Potter, but in a shorter form, so I hope some of you will be pleased to have a fuller version.

After a little time, the door opened a few inches and her face peeped out. Seeing us, she opened it and said 'Cum in …' To my astonishment and delight I saw she was wearing a knitted tea cosy, like a fitting cap. One of those old-fashioned ones with an opening for the handle and a smaller one for the spout. It was extremely fetching! We followed her in. Silver mounted guns of the past hung over the doors. It was very quiet, but in an 'alive' way. The living room was well lit with windows and warm with a good coal fire. She settled in her armchair and we 'sat on the edge of our coppy-stools' like Pigling and Pigwig in Mr Mcgregor's house. But she soon made us feel at home and talked mostly to Delmar about his painting the fells from high level.

She seemed easily amused, and when she laughed she rolled a little and slapped her knee, with her small chubby hands. We saw her as alert and lively as one of her own little Animal characters. Later on, she knew this and was pleased – see her sketch of herself as Tiggy-winkle for me at the end of a letter – and when she removed her tea cosy; behold a charming black velvet ribbon over her silver hair with a little bow on top ... and she knew we thought she was fetching – and chuckled. So far as I can tell from the books about her, not many had this intimate view. Those of you who already love her from reading her Diary, and her biographies and letters to friends but especially from her books, would have been as charmed as we were. But also it was the stability and down-to-earth strength of her character that dominated with such gentle effect – one knew one was in the presence of one of the GREAT Natures of our land.

Delmar was thrilled to be allowed to see her original paintings (4 times as large as in the books) and to be told a lot about their reproduction – certain colours 'taking' and others not – especially blue. She bought two of his large watercolours of the fells, one in Winter with a Fox. 'Not that I like it best, but you will never do the same again.' You will see what she says about trees in a letter that was most helpful to him.

How brave she and William were to persist in getting married! Her nephew John Heelis recently told me that the Potters objected to William being 'in law', and the Heelis's, to the Potters being 'in trade'! They, particularly her father, made it as difficult as possible, in a way very like the possessive mania of the father in the play 'The Barretts of Wimpole Street'.

Beatrix had so much affection for her brother and admiration for his Oil Paintings. In one room she had several landscapes in frames and spent quite a time for me to see them. I thought his painting of woodland was really poetic, though tinged with sadness.

Once, when we were walking in the garden I saw the kind of light chicken-wire made cage – that when a patch of grass has been nibbled you move it on to the next. She said 'I keep Rabbits because when children are brought to see me, they always think that Peter Rabbit lives with me – and expect to see him'. When they ask 'Is one of them Peter?' I reply 'No – but a near relation'. Would you believe the lie that Beatrix Potter disliked children – that she used to chase children away from her garden? It was put about by a neighbour's child who was the ringleader of a group who used to come in the backway and disturb Beatrix with milling around and shouting when she was trying to write quietly. She did not 'chase' them – she came out and said 'You have your own gardens to play in and the lane – you must keep out of my garden! – and I need to live in peace'. She told me this herself and pointed to the gate they came in by from the lane.

Another woman also put out lies about Beatrix. She was jealous because her husband was ready to do his best in the jobs on the farm or garden – was happy to be working there. She couldn't stand it he hurried when Mrs Heelis called him! Twisted and acid with jealousy, were both these women.

How glad I am that Beatrix made a lot of money from her books, and could be independent! If you have ever suffered from slander you will know that such things as a bit of money have their use for getting away from it, or having a change! You also find out who your friends are. John Heelis had never met his aunt Beatrix – and he asked 'Did she really dislike children? Oh John – how could you ask such a thing! Everything she did or wrote denies it! Just remember that Jesus was accused of healing by the power of Satan.

Well, we were walking in the garden of Castle Cottage … she said something about 'zig-zag clover' and I had to admit I'd never seen it. 'I'll show you some'. On we went to where the old apple trees made a small orchard, and in the grass on the sunny side were several rows of wild flowers she had collected and planted herself. 'It's important for people to know and to preserve them.' She picked a stalk of zig-zag clover and gave it to me, keeping a sprig for herself.

When I took my leave, we both knew we might never meet again … These intimations are natural when there is affection and a likeness in outlook and disposition – and she pulled my face down to hers and kissed me. I couldn't say anything, not even goodbye … and walked slowly away. It affected me so much to see how tiny she was by then.

Near the gate, I looked round. She was standing just where Timmy Willie stood when he was waving goodbye to Tommy Town Mouse – and she waved her zig-zag clover leaf just as he did – knowing that I would see all and understand. …

But what a character was there. She was not one to evade awkward corners by hiding in a haze of politeness. Her new neighbours came and had settled into their homes. It was still the old-fashioned way to 'call' upon them. She did 'call' – but wearing her old clothes, and knocked at the back door. If – as once or twice happened – she was turned away, she knew what they were like, and did not return! Once, when the weather was bad and she was out in the lane wearing a sack over her back and collecting fallen acorns, a Tramp came

The Bield, Little Langdale

by and said 'It's a sad time for the likes of Thee and me!' This story has been told before but not everyone knows it, so I include it here.

I (now 98) am old enough to have seen many Tramps and once I came upon a Woman Tramp who looked very old. She was sitting with her belongings around her, in a sheltered hollow – smoking a pipe! she fitted in so naturally to that wild and brambled place, that I felt no astonishment. Yet in those days it was considered quite avant garde for a woman to be seen smoking a cigarette out of doors. The Artist Arthur Rackam should have been there to have recorded that picture. When I wrote that word 'astonishment' I remembered a delightful Brazilian lady – a close friend – whose English was always refreshing, saying 'I was etonated!' – which seemed so expressive.

This 'doodling in the mind' has no business in a proper book – but why worry? I write, as sparrows pick up crumbs – but 'Sister' (sees't thou) as a responsible owner for years of a fine fell Sheep Farm (The Grove at Hartsop) and member of the Women Farmer and Garden Association, ready for a joust with the Milk Marketing Board and knowing the far side of a sheep from the near I can be a bit eccentric!

Another one who fell as we did under the spell of Beatrix Potter's naughty charm, and beguiling unexpectedness, was a professor from Durham

University. We met him when staying at the Woolpack Inn (another doodle but if I don't put it in here it will get lost) which, by the way also runs a flock of Herdwicks on the Slight Side of Scawfell. I think the reason why they always got a first prize in Sheep Shows in my day was because their heaf (nibble grounds) is so high up that when their fleeces get soaked with rain, it is the sea wind that dries them and makes them so fine and fluffy!) He told us that out of the answers to a questionnaire about 'Diseases in Hill sheep' the best came from a Farmer in Sawrey – and he decided that he must meet him – So he wrote and date was made. As we did, he found the door and knocked – but was let in by a thin old Maid dressed in the stiff black and white starched costume of an Edwardian Parlourmaid. With polite disapproval she led the way upstairs and opened the door of a small bedroom. Cosily wrapped up in a large bed sat a diminutive old lady! A famous talk followed to the satisfaction of both – and Tea was brought in. Farmer Heelis knew her subject …

It was in that same bed that when recovering from an illness she told me that looking out of the window at the vegetable garden every day, one solitary cabbage remained, though she kept asking for its removal. 'In the end I was sick of seeing it. I got out of bed, put on Willie's old mac, crept out and pulled it up' (said with a naughty satisfaction). She also told me how sad she was when one summer ago she had lost her Engagement Ring in the hayfield. All those lonely years she had mourned the death of young Warne while they were still secretly betrothed. I think the large quiet and distinguished Country Lawyer was the one man who could have given her the happiness and comfort of a happy married life. It was a great privilege to have been so kindly taken into their home and safe retreat. After she died, and I was on my own, I felt her very near me.

But although on the whole a cheerful person, why should I feel so sad at this moment of remembrance? It is the strong visual impression of the eyes of a child, exhausted by posing for long sessions for the many photographs taken of her. At that time photography was a lengthy business for each pose. Her eyes show it –– but far more … a deep hurt and revulsion, never told – not even in her secret diary in code. Only – 'how much longer' could she stand it? I know (I suffered things in childhood I could not stand now, so I understand).

Having got so far, have you perhaps wondered what relation the title had to do with anything yet? Delmar used to say that beyond the tales and their pictures and episodes of country life was a clear perception of relationships

Early sketch of Jemima Puddleduck by Beatrix Potter (Private Collection)

as would help children to cope with people and situations later in life as they came to it. Characters in the Pooh books also play their part in the same unsentimental way.

What I can only call Whispers in the Grass is the underlying poetry of the words that in themselves and what is still left unsaid or seen, on the imagination to be led of a further continuation of events after the end of the story or left out of a picture. for instance many artists would have been tempted to show the errant Jemima Puddleduck flying over fields farmlands and woodlands, but Beatrix gives a sense of height and movement over a land too large to paint. The Countess of Huntingdon and Jack her husband came to stay at the Bield so as to have two evening readings of her book with us before showing it to Publishers.

It was the first on Beatrix Potter and in some way the best Margaret had perceived this hidden quality of poetic mystery in *The Tale of Mrs Tittlemouse*, 'and she polished her little tin spoons', the picture suggesting the whole story of an Innocent little life full of its own definite character. There is a luminosity about the eyes that one sees in some of the finest photographs of Mice (particularly the Dor). Fra Angelico gave his child like Saints ... the same light of an innocent and wondering spirit. In these secret but revealing glimpses lie her poetic gifts.

THE grass is cut, the flowers fade,
The Sun dries up the hay;
Was all this beauty only made
To perish in a day?

The Infant lies with ne'er a sheet
Nor coverlet of white;
But in the manger, Ah, how sweet
The dried hay smells tonight!

J. DE V.

∴

From Delmar & Josephine Banner
The Bield, Little Langdale, Ambleside

Christmas card by Josefina de Vasconcellos

II

ONE CHRISTMAS I sent her a card of a verse of ours and one of my wood-cuts.

"Your poem made me cry".

Beloved – your "Poems" dried my tears.

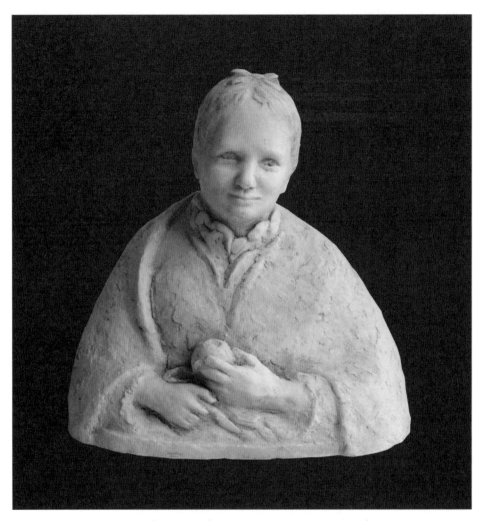

Peeling apples to make a pie for Willie

by Josefina de Vasconcellos
(cast in J B formula)

12

Letters from Beatrix Potter
to Delmar Banner and 'Pigwig'

with drawings by Beatrix Potter
from The Armitt Trust

Castle Cottage
Sawrey

Oct 11. 36

Dear Mr Banner,

. . . If you and Mrs Banner
don't mind the trouble of coming
here it would be a relief —
I would undertake to move you
one way in the car if you could
come to lunch on Thursday, & I
could pick you up in Coniston
about 12·45. and we should be very
glad to see you — and not sorry
if health resort has proved propitious.

Castle Cottage
Sawrey

Oct 11. [19]36

Dear Mr Banner,

 If you and Mrs Banner don't mind the trouble of coming
here it would be a relief – I would undertake to move you one
way in the car if you could come to lunch on Thursday – I could
pick you up in Coniston about 12.45 and we should be very glad
to see you – and not sorry if Heathwaite has proved propitious.

There seems to be a good deal of
competition, not by despoilers, but by
will meaning outsiders who may run
prices up through mis directed zeal.
at Ellwwths

Yrs sincerely

H. B. Kulir

I can explain better when I see you.

It is a muddle. I have the deepest
respect and admiration for the National
Trust as an institution; but the present
officials are not very satisfying at buying
properties.   unsatisfactory.   I don't
want them meddling in Little Langdale.

There seems to be a good deal of competition at Elterwater, not by despoilers, but by well-meaning outsiders who may run prices up through misdirected zeal.

yrs sincerely

H.B. Heelis

I can explain better when I see you. It is a muddle. I have the deepest respect and admiration for the National Trust as an institution; but the present officials are not very satisfying at buying properties. Unsatisfactory. I don't want them meddling in Little Langdale.

Castle Cottage
Sawrey

Oct 13.36

Dear Mr. Banner

I am sorry to have to tell
you that you have no chance
of buying the Brow, thanks to the
proceedings of the supporters of the
National Trust. "Friends" with
money have come forward to bid
for various lots, including Dale
End farm which will include the
Brow. I think I am to
be let alone (so far as the
amenity tribe are concerned); but

Castle Cottage
Sawrey

Oct 13. [19]36

Dear Mr Banner,

    I am sorry to have to tell you that you have no chance of buying the Brow, thanks to the proceedings of the supporters of the National Trust. "Friends" with money have come forward to bid for various lots, including Dale End farm which will include the Brow. I think I am to be let alone (so far as the amenity tribe are concerned); but

only
upon a tacit understanding that
I shall not **sell** for building.
My hands are tied.

Mr Heelis says that under the
circumstances it would be hopeless
for you to bid.

I wish that some likely place may
turn up at Hawthwaite to make
amends. There may be some
advantages ~~of~~ in settling outside the
area of a potential "National Park".
I have never been able to fathom what
exactly its advocates are aiming at; but
I am sure it means interference with
other peoples property and freedom of
buying and selling. And probably an
over visited show place in another 20 years;
it grows more crowded every season.
I remain yrs sincerely
H B Heelis

only upon a tacit understanding that I shall not sell for building. My hands are tied.

Mr Heelis says that under the circumstances it would be hopeless for you to bid.

I wish that some likely place may turn up at Heathwaite to make amends. There may be some advantages in settling outside the area of a potential "National Park". I have never been able to fathom what exactly its advocates are aiming at; but I am sure it means interference with other people's property and freedom of buying and selling. And probably an over-visited showplace in another 20 years; it grows more crowded every season.

I remain yrs sincerely

H.B. Heelis

Sept 8. 36

Castle Cottage
Sawrey
nr Ambleside

Dear Mr Banner,

There are no particulars available yet as to how the Elterwater Hall estate is going to be lotted. The announcement ought to be in the Gazette amongst the auction advertisements if it is really going to be sold in October. I know that some of the Elterwater cottagers are very uneasy about their homes. Some of the land is below the Colwith Bridge – Elterwater road and Elterwater tarn; presumably that will be put up with the Hall. There is an outlying farm in great Langdale, and another farm looking into Little Langdale. If you look at the map you will see there is a road going over from Little Langdale by Dale End to Birk Hill. On the actual signpost is the delightful direction to "Hodlet's Nest" but I confess I have never been over it, it turns off before you go down to little Langdale p. office. Mercifully it is not a good car road, and I am not one of those thoughtless old bodies who demand charabanc routes because they cannot hope to walk. But without personally knowing the place I think there must be a very fine view from the Elterwater land up that road. Hacket has never belonged to the Robinsons; it was Church land formerly. I mention it because it is a corner of the triangle of land. The eastern slope is attractive copse wood land. The chances are it will be sold in large lots and broken up later. I do hope

[44]

                                             Castle Cottage
                                             Sawrey
                                             Nr Ambleside

Sept 8th [19]36

Dear Mr Banner

There are no particulars available yet as to how the
Elterwater Hall estate is going to be lotted. The announcement
ought to be in the Gazette amongst the auction advertisements if
it is really going to be sold in October. I know that some of the
Elterwater cottagers are very uneasy about their homes. Some
of the land is below the Colwith Bridge–Elterwater road and
Elterwater tarn; presumably that will be put up with the Hall.
There is an outlying farm in Great Langdale, and another farm
looking into Little Langdale. If you look at the map you will see
there is a road going over from Little Langdale to by Dale End
to Brid Hill. On the actual signpost is the delightful direction to
"Hoolet's Hut" but I confess I have never been over it. It turns
off before you go down to Little Langdale p. office. Mercifully it
is not a good car road, and I am not one of those thoughtless old
bodies who demand charabanc routes because they cannot hope
to walk. But without personally knowing the place I think there
must be a very fine view from the Robinson Elterwater land up
that road. Hackett has never belonged to the Robinsons; it was
church land formerly. I mentioned H^t. because it is a corner of
the triangle of land between roads. The eastern slope is attractive
copse woodland. The chances are it will be sold in large lots and
broken up later – I do hope

that good class houses — like the modern (though slightly bare — *unpicturesque* Hackett) will get built above the Hall, if there is building. The view *on the eastern slope* would suit you — panoramic, not grandeur.

You have one advantage in your quest, unlike the majority you prefer to be off the road and high up, which should make for an easier price per acre. I think it is doubtful if there will be small lots, but it would be quite worth your while to walk up that road if you are in Langdale. I think there must be a fine view into the back of Wetherlam and Greenburn. I understand well what you mean. We are a little too near the village here during August and Bank holidays — But on the other hand it is wise not to be too remote, for a permanent home.

'Pig-wig' — bless her — is still at the ecstatic stage when she would like to be over the hills and far away — but what about carting coal if there is no road?! What about occasional daily help? *not to mention large canvasses and works of marble — a green slate? Try a query!!* The best situation is with your back to a hamlet, out of ear shot, with a safe uninterrupted view in front — I fancy there might be cheaper land between Coniston-Tower-Heathwaite than anywhere in the Langdales. but I will certainly inquire, no trouble as I will be going up shortly — It depends how it is lotted.

I very much enjoyed the "private view" I was a little alarmed to find how seriously you had taken my criticism — But they are wonderfully better — breadth, atmosphere; without sacrifice of colour and drawing! I remain with kind regards to you both yrs sincerely
Beatrix Heelis

that good class houses – like the modern (though slightly conspicuously like Hackett) will get built on the eastern slope above the Hall, if there is building. The view would not suit you – panoramic, not grandeur. You have one advantage in your quest, unlike the majority you prefer to be off the road and high up, which should make for an easier price per acre. I think it is doubtful if there will be small lots, but it would be quite worth your while to walk up that road if you can in Langdale. I think there must be a fine view into the back of Wetherlam and Greenburn. I understand well what you mean – We are a little too near the village here during August and Bank holidays – But on the other hand it is wise not to be too remote, for a permanent home. 'Pigwig' – bless her – is still at the ecstatic stage when she would like to be over the hills and far away – but what about carting coal if there is no road?! What about occasional daily help? not to mention large canvasses and blocks of marble – and green slate? Buy a quarry?? The best situation is with your back to a hamlet, out of ear shot, with a safe uninterrupted view in front. I fancy there might be cheaper land between Coniston-Town-Heaththwaite than anywhere in the Langdales, but I will certainly inquire, no trouble as I will be going up shortly – It depends how it is lotted.

I very much enjoyed the 'private view'. I was a little alarmed to find how seriously you had taken my criticism. But they are wonderfully better – breadth, atmosphere; without sacrifice of colour and drawing. I remain with kind regards to you both yrs sincerely

Beatrix Heelis

I paid /1400 for the Busk farm in the same valley. I think Dale End has rather more land and better meadows; but less road frontage, and in worse condition. That is how I arrive at a guess at price.

It is worth more than the Busk. ... Except for us lunatics who spend money on a View, it is in moderate condition; no lack of water.

Give my sincere sympathy to your wife.

Yours sincerely H B Bielces

Perhaps you would not be selling that Coniston view-on-paper before I see you again. A day will come when my old legs refuse to climb stiles and wet lanes, so painted views will have to suffice.

[Sept 14th 36]

I paid £1400 for the Busk farm in the same valley. I think Dale End has rather more land and better meadows; but less road frontage, and in worse condition. That is how I arrive at a guess at price. It is worth more than the Busk – except for us lunatics who spend money on a View, it is moderate condition, no lack of water. Give my sincere sympathy to your wife.

Yours sincerely
B. Heelis

Perhaps you would not be setting that Coniston View-on-paper before I see you again. A day will come when my old legs will refuse to climb stairs and wet lanes, so painted views will have to suffice.

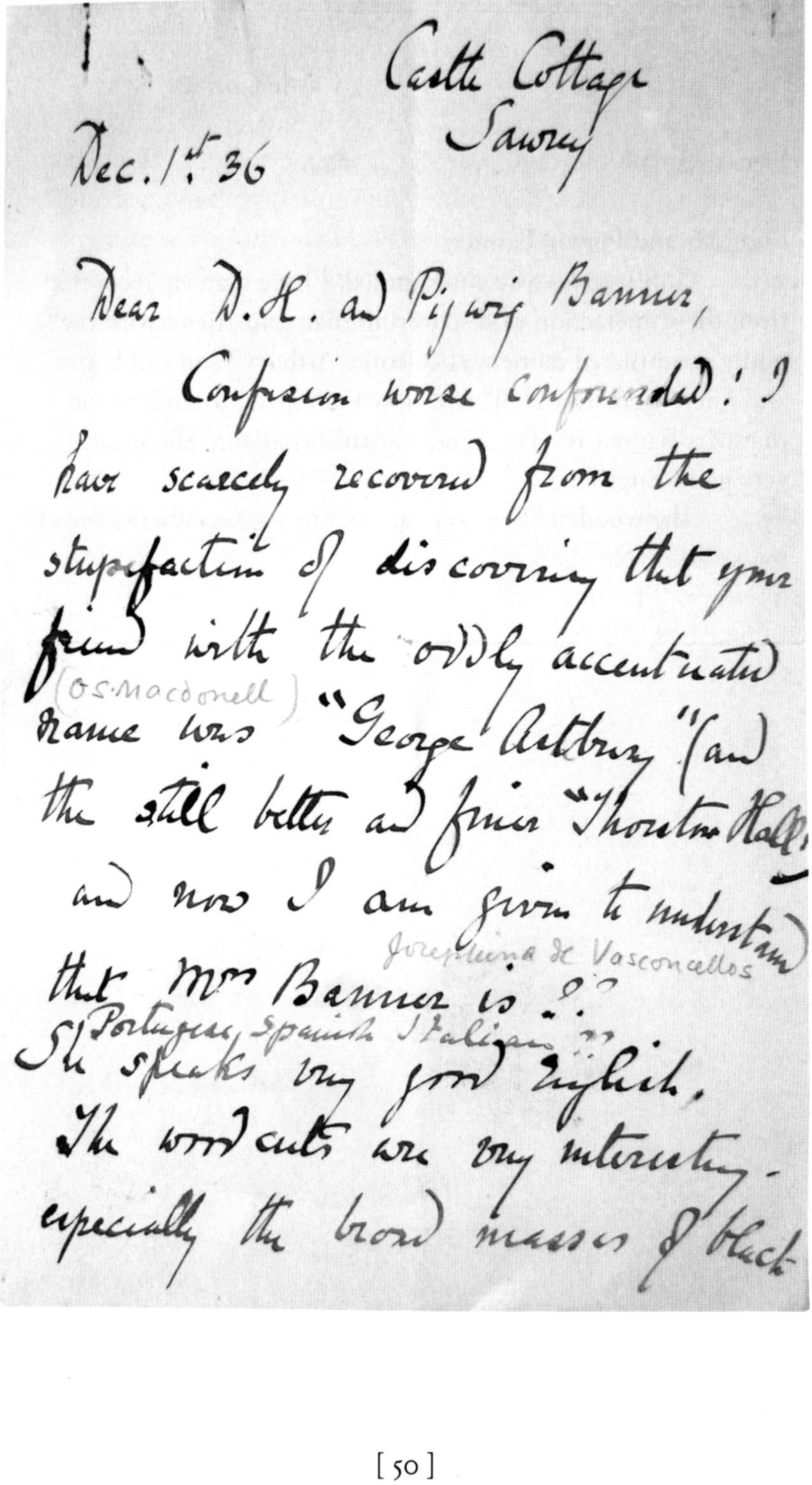

Castle Cottage
Sawrey

Dec. 1ˢᵗ 36

Dear D.H. and Pigwig Banner,

Confusion worse confounded! I have scarcely recovered from the stupefaction of discovering that your friend with the oddly accentuated name was "George Astbury" (and the still better and finer "Thornton Hall"

(O.S. Macdonell)

and now I am given to understand that Mrs Banner is ?

Josephina de Vasconcellos

She speaks very good English.

[Portuguese, Spanish, Italian]

The wood cuts are very interesting, especially the broad masses of black

Castle Cottage
Sawrey

Dec. 1ˢᵗ [19]36

Dear Mr and Pigwig Banner,
    Confusion worse confounded! I have scarcely recovered
from the stupefaction of discovering that your friend with the
oddly accentuated name was "George Astbury" and still better
and finer "Thorston Hall" and now I am given to understand
that Mrs Banner is?? Portuguese Spanish Italian?? She speaks
very good English.
    The woodcuts are very interesting – especially the broad
masses of black

as in 26. But its cruel to try
to deal with the sun & sky in
scratches — it spoils 46, also 39.
8.22 & the frontispiece are the
most satisfying. I am wroth with
"Fanny Johnson" for setting "Josephine"
an impossible task. Had her
poems been equal to Blake's
they should like his have been
illustrated in colour — or shaded into.
Some of the more simply expressed
pieces are beautiful; reflected beauty.
There is nothing new under the sun.
    I think Pigwig — Josephine
is a very interesting artist — I live
in hope of meeting her binded
cat one day — which is more on my

as in p. 26. But its cruel to try and deal with the sea and sky in scratches – it spoils 46, also 32. 8, 22 and the frontispiece are the most satisfying. I am wroth with "Fanny Johnson" for setting "Josephina" an im-possible task. Had her poems been equal to Blake's they should like his have been illustrated in colour – or shaded ink. Some of the more simply expressed pieces are beautiful; reflected beauty. There is nothing new under the sun.

I think Pigwig–Josephina is a very interesting artist – I live in hope of meeting her blinded cat one day – which is more on my

level! I hope eventually these
will be found a "site" — nothing
suitable has yet been heard of.
Elterwater Hall is sold to a Mr
Hodge from Southport £1675 — "Thrang"
a cottage near Chapel Stile has been
sold for £750 with 9 acres, but I did
not write to you about it, as it was
close to gunpowder-works-garden-city,
and the school — too low down, and
too "thrang". As "thrang as inkle weavers"
as the saying is. It takes time.
There were some lovely days,
above the fog, at the end of last week.
On Saturday I was looking over from

level! I hope eventually there will be found a "site" – nothing suitable has yet been heard of. Elterwater Hall is sold to a Mr Hodge from Southport – "Thrang" a cottage near Chapel Stile has been sold for £750 with 9 acres, but I did not write to you about it, as it was close to gunpowder works, garden city and the school – too low down, and too "thrang". As "thrang as intile weavers" as the saying is. It takes time.

There were some lovely days, above the fog, at the end of last week. On Saturday I was looking over from

the road above Elterwater hall, over Dale End, with a frosty sun set behind Wetherlam — gradually fading into dark. and a great moon rising over Kirkstone. The weather has gone again, wind and sleet.

I like your water colour extremely; it grows upon us.

And I like the cover of the poems. and p. 38:9. Sorry I cannot honestly admire all of it, it is too fantastic, a little forced. Blake was naturally mad so to speak. I must stop now for post — with kind regards to you both yrs sincerely

Beatrix Heelis

But assured I will continue to look out. and I have not scheduled the lonning, but any light house would be discovered up above. as was plain on Saturday night.

the road above Elterwater Hall, over Dale End, with a frosty sunset behind Wetherlam – gradually fading into dark – and a great moon rising over Kirkstone. The weather has gone again, wind and sleet.

I like the cover of the poems and p. 38-9. Sorry I cannot honestly admire all of it, a little forced. Blake was naturally mad so to speak. I must stop now for post – with kind regards to you both yrs sincerely

Beatrix Heelis

Rest assured I will continue to look out, and I have not scheduled the lower land, but any large house would be desecration up above, as was plain on Saturday night. You cannot have a very small one on acct of the studio. At Kearstwick all is on a larger scale though no longer all. Patience!

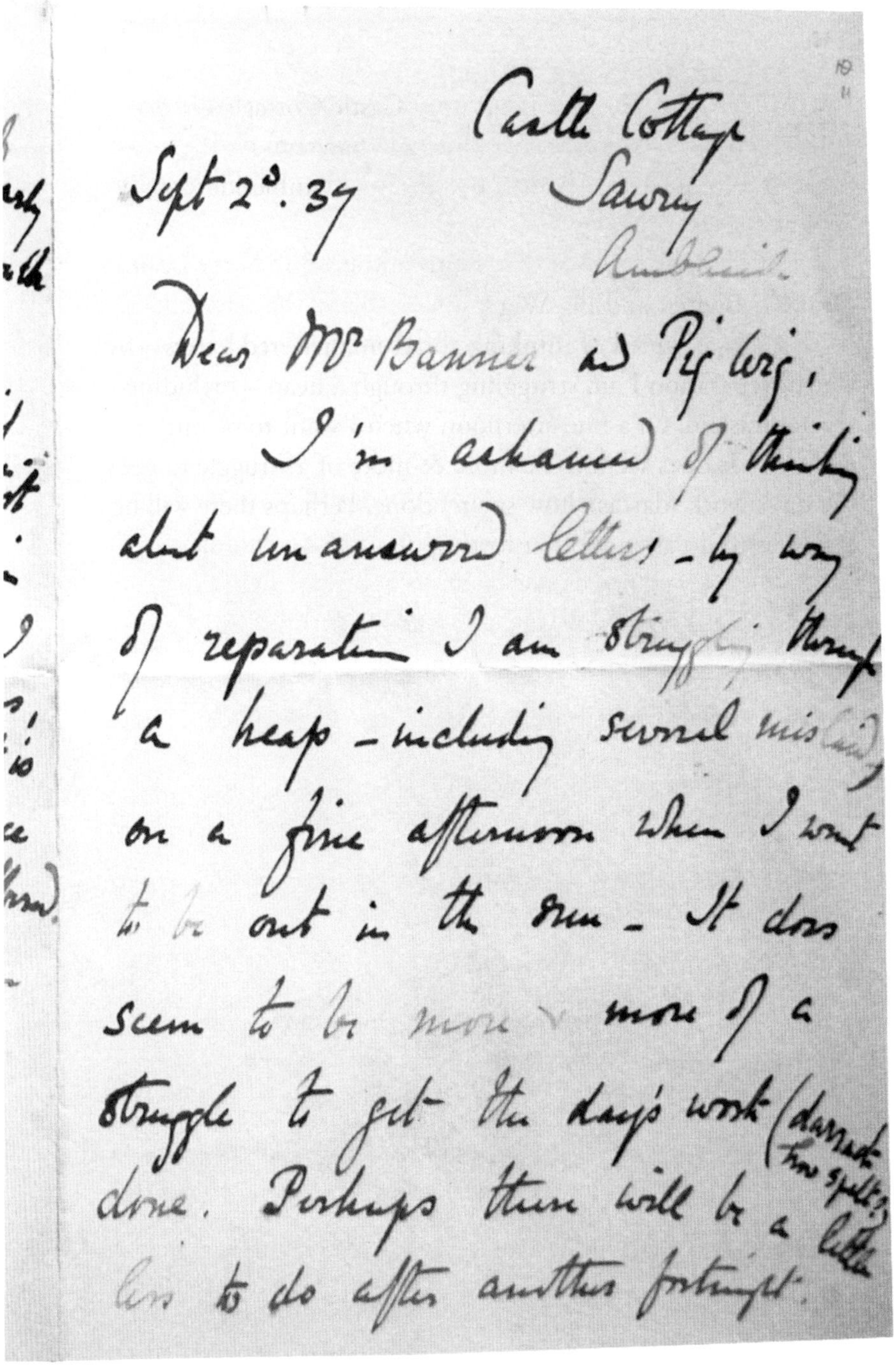

Castle Cottage
Sawrey
Ambleside

Sept 2º. 39

Dear Mr Banner and Pig wig,

I'm ashamed of thinking about un answered letters – by way of reparation I am struggling through a heap – including several mis(laid) on a fine afternoon when I want to be out in the sun – It does seem to be more & more of a struggle to get the day's work (damn me how spelt) done. Perhaps there will be a little less to do after another fortnight.

Castle Cottage

Sawrey

Ambleside

Sept 2nd 37

Dear Mr Banner and Pig Wig,

    I'm ashamed of thinking about unanswered letters – by way of reparation I am struggling through a heap – including several mislaid, on a fine afternoon when I want to be out in the sun – It does seem to be more & more of a struggle to get the day's work (darrask how spelt?) done. Perhaps there will be a little less to do after another fortnight.

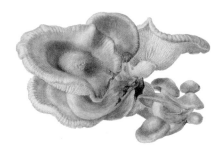

much of it is interesting – even dreans.
and deficiencies of shippons. I have
been up occasionally to Dale End; I
expect a wigging next visit; as I
hear the masons have uprooted 2
old rose trees and the box hedge which
smothered the poor old Mrs Birketts
doors and windows. They will grow
again.

I'm not too proud to come
up in a milk float provided
you can haul me into it!
I was at Coniston this morning and
looked up towards Heathwaite and

Much of it is interesting – even drains and deficiencies of shippons. I have been up occasionally to Dale End; I expect a wigging next visit, as I hear the masons have uprooted 2 old rose trees and the box hedge which smothered the poor old Mrs Birkett's doors and windows. They will grow again.

I'm not too proud to come up in a milk float provided you can haul me into it! I was at Coniston this morning and looked up towards Leathwaite and

the lights & shadows in Minus
Ghyll — it is rather beautiful.
weather? The drains were
"making a lot of water" & requiring
still more drain tiles.
I also thought of Pig wig(?) &
the Catacombs — incongruously —
though not disconnectedly — for I
have been persuading — or trying to persuade
a friend to have some relics buried
in holy earth. I do not like
the early Christians in pieces in
modern surroundings. And if I could

the lights and shadows in Mines Ghyll – it is rather beautiful weather – The drains were "making a lot of water" and requiring still more drain tiles.

I also thought of Pigwig and the Catacombs – incongruously – though not disconnectedly – for I have been persuading – or trying to persuade – a friend of mine to have some relics buried in holy earth. I do not like the early Christians in pieces in modern surroundings. And if I could

explain myself – I doubt if Pi arg
in all earnestness can recapture the touching simplicity of early
Christian art. The nearest approach
would be from the outlook of a little
child, or an unsophisticated nature (if
such still exists). For a trained artist
equipped with tools & modern brushes & pencils
to convey by scratches. I maintain
is <u>imitative</u>, not spontaneous. But I
apologize for my criticism of the poems,
I thought they were written by a girl; it is
touching to think that they are utterance
of one older who has felt & suffered.
I hope very much to meet when
you come to Heathwaite –
yrs sincerely
Beatrix Heelis

explain myself – I doubt if Pigwig in all earnestness can recapture the touching simplicity of early Christian art. The nearest approach would be from the outlook of a little child, or an unsophisticated nature (if such still exists). For a trained artist equipped with tools and modern brushes & pencils to convey by scratches I maintain is *imitative*, not spontaneous. But I apologize for my criticism of the poems, I thought they were written by a girl; it is touching to think that they are an utterance of one older who has felt and suffered. I hope very much to meet when you come to Heathwaite –

Yours sincerely

Beatrix Heelis

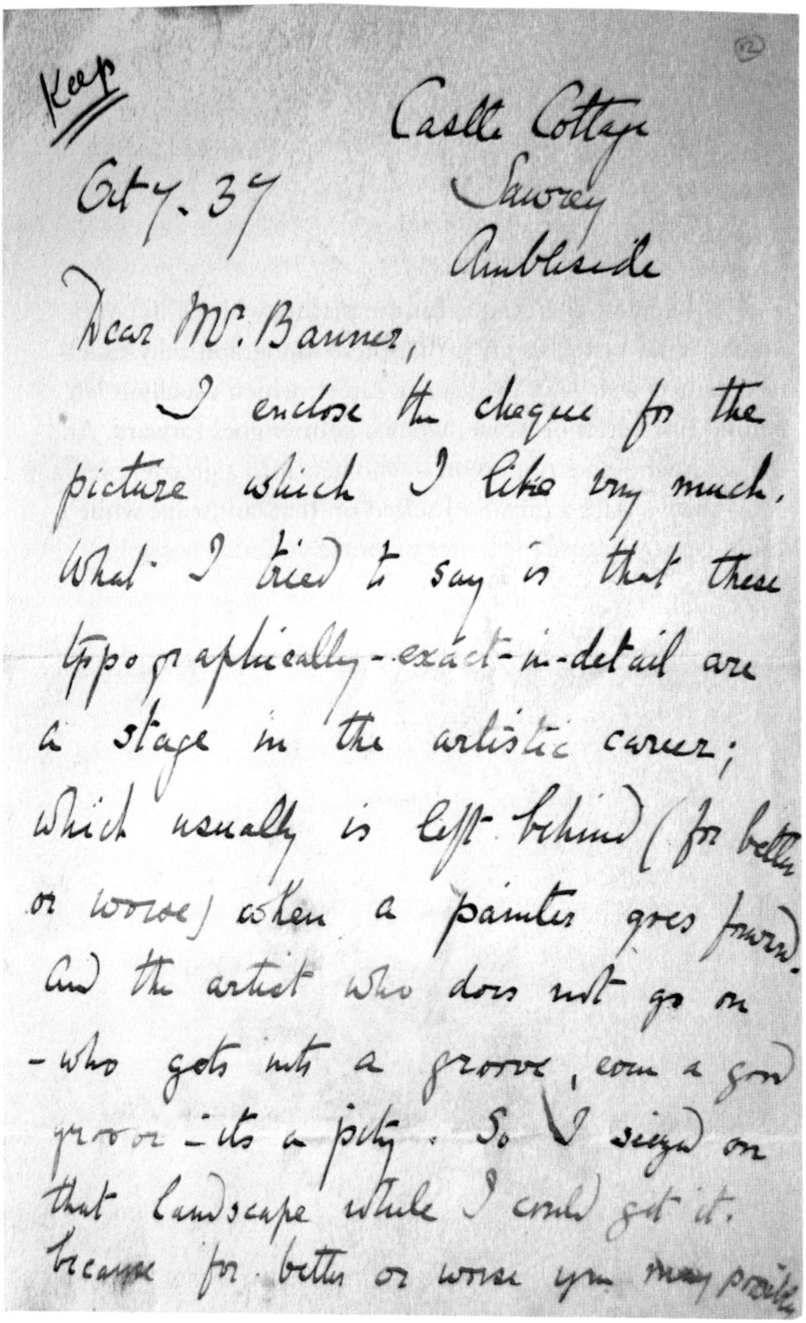

Keep //

Oct 7. 37

Castle Cottage
Sawrey
Ambleside

Dear Mr. Banner,

I enclose the cheque for the
picture which I like very much.
What I tried to say is that these
typographically - exact - in - detail are
a stage in the artistic career;
which usually is left behind (for better
or worse) when a painter goes forward.
And the artist who does not go on
- who gets into a groove, comes a good
proverb — it's a pity. So I seized on
that landscape while I could get it.
because for better or worse you may probably

Castle Cottage
Sawrey
Ambleside

Oct 7 [19]37

Dear Mr Banner,

    I enclose the cheque for the picture which I like very much. What I tried to say is that these topographically-exact-in-detail are a stage in the artistic career; which usually is left behind (for better or worse) when a painter goes forward. And the artist who does not go on – who gets into a groove, even a good groove – it's a pity. So I seized on that landscape while I could get it, because for better or worse you may possibly

not paint like that in a few months or years hence. The clouds were wonderful in some of the landscapes. Only in one or two had you found the heartlessness to let the mist come down over your beloved detail of fell and screes and rock.

Perhaps some time if you are at a loose end in Surrey exile you might consider studying trees. You don't care to choose landscape low enough down to require much appreciation, but it is useful to understand them. And incredible how badly many professional landscape painters don't. I mean they have never considered how the branches not paint like that in a few months or years hence. The clouds were wonderful in some of the landscapes. Only in one or two cases had you let the mist come down on your beloved detail of fell and screes and rocks. Perhaps sometime if you are at a loose end in Sawrey exile you might consider studying trees having mastered the form of clouds. You don't care to choose landscape low enough down to require much appreciation but it is useful to understand them, and incredible how badly many professed woodland landscape painters don't. I mean they have never considered how the branches

grow from a tree trunk. For instance the ash 'igdrasil' the tree of heaven. Every year a new short [...] if you study an ash [...] you will see every branch from the main trunk, or the stem of the young sapling, has come out in curves; and curved on & on with the weight of foliage., We can tell every tree in winter without reference to foliage, by its mode of growth. So study them some spare moments M⁵ Barimer; they will repay — they are — in the right place — as beautiful as rocks. And they have a nobility of growth

other species in contrast grow upward. If

which is naturally indicated over & above

grow from a tree trunk. For instance the ash 'igdrasil' the tree
of heaven. Every year a new shoot [sketch] If you study an ash
[sketch] you will see every branch from the main trunk, or the
stem of the young sapling, has come out in curves; and curved
on and on with the weight of foliage. Other species in contrast
grow upward. [sketch] We can tell every tree in winter without
reference to foliage, by its mode of growth. So study them some
spare moments Mr Banner; they will repay – they are – in the
right place – as beautiful as rocks. And they have a nobility of
growth which is usually overlooked.

I had a most enjoyable visit, thanks
to you and Josephine — and — Peggy;
It is a dear place, and good people.
The puppies were 'ticing. But young
dogs require much exercise; I must
wait till another of our working collies
is ready to return to replace old Fly.
Now I have a tiresome committee
meeting to go to, so won't inflict any
more preaching upon you — or Pig, why
She certainly took my comments on
the rays of sunshine like an angel!
I will go some time to Loston and
look at old Ned's grave.. I had

(Ned Nelson — carved by J B)

lots of things I meant to talk about
but was feeling muzzy. What a
pleasant house and float and charioteer
Another time when you are at Cousten I want to
show you rather a fine old farm house.

I had a most enjoyable visit, thanks to you and Josephine-and-Piggy! It is a dear place, and good people. The puppies were 'ticky. But young dogs require much exercise; I must wait till another of our working colleys is ready to return to replace old Fly. Now I have a tiresome committee meeting to go to, so won't inflict any more preaching upon you – or Pigwig. She certainly took my comments on the rays of sunshine like an angel. I will go sometime to Lorton and look at old Ned's grave. I had lots of things I meant to talk about but was feeling muzzy. What a pleasant horse and float and charioteer. Another time when you are at Coniston I want to show you rather a fine old Georgian house. It has no water supply, and sundry drawbacks. It is not vacant. Sitters at present. A curious old place.

Yrs sincerely

Beatrix Heelis

I think probably in the end I will have it pulled down, when the present occupiers is dead.

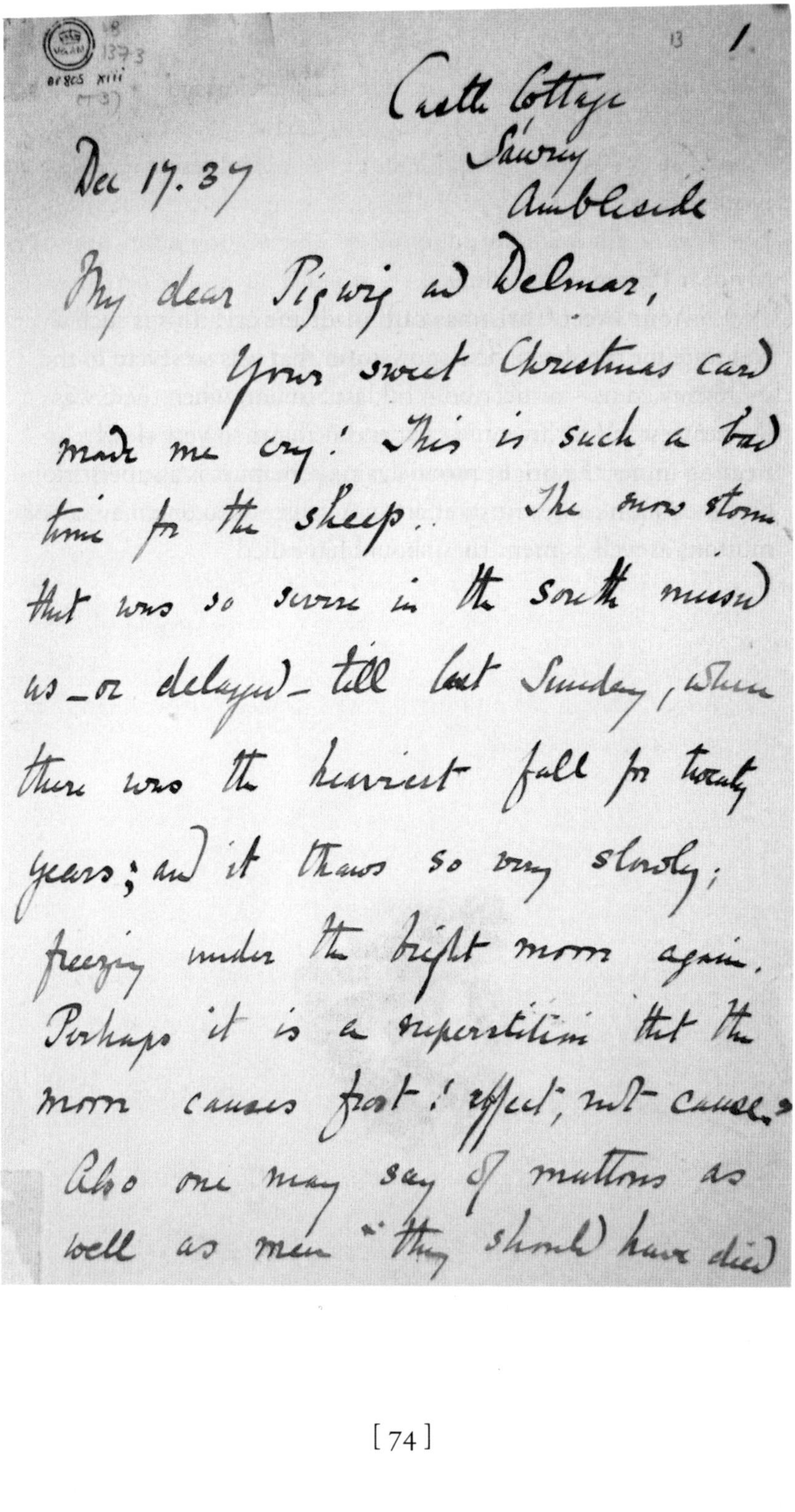

13   1.

Castle Cottage
Sawrey
Ambleside

Dec 17. 37

My dear Pigwig and Delmar,

Your sweet Christmas card
made me cry! This is such a bad
time for the sheep.. The snow storm
that was so severe in the south missed
us — or delayed — till last Sunday, when
there was the heaviest fall for twenty
years; and it thaws so very slowly;
freezing under the bright moon again.
Perhaps it is a superstition that the
moon causes frost! effect; not cause.
Also one may say of muttons as
well as men "they should have died

Castle Cottage

Sawrey

Ambleside

Dec 17 [19]37

My dear Pigwig and Delmar

     Your sweet Christmas card made me cry! This is such a bad time for the sheep. The snowstorm that was so severe in the south missed us – or delayed – till last Sunday, when there was the heaviest fall for twenty years; and it thaws so very slowly; freezing under the bright moon again. Perhaps it is a superstition that the moon causes frost; effect, not cause? Also one may say of muttons as well as men "they should have died

here after"; but its ghastly to think about
the carrion crows.. We feed them
here; but they don't thrive on hay,
and its impossible to feed a big fell
flock.. Old customs become disused.
I have just been telling the men to
cut some "croppit ashes". That is
why you see the surviving croppit
ashes near fell farm-houses. It used
to be the custom to lop ashes and hollies
for the sheep. Two quarry men
had an adventure on Tuesday coming
down from work at the Old man quarries
they were caught by an avalanche - One
got clear & fetched help, the other lad
was found at 2 a.m. he had got
his head out and was shouting.." Quite unhurt.
after 9 hours buried.

hereafter"; but its ghastly to think about the carrion crows. We feed them here, but they don't thrive on hay, and its impossible to feed a big fell flock. Old customs become disused. I have just been telling the men to cut some "croppit ashes". That is why you see the surviving croppit ashes near fell farm-houses. It used to be the custom to lop ahes and hollies for the sheep.

Two quarry men had an adventure on Tuesday coming down from work at the Old Man quarries they were caught by an avalanche – One got clear & fetched help, the other lad was found at 2 a.m. he had got his head out and was shouting!! Quite unhurt after nine hours buried.

so the landscape is very old fashioned Christmas – we keep well here –

I am sending you Christopher le Henry's music, it was very charming when he played it but I think too difficult for children. I look regularly at advertisements but in vain – I think one ~~was with~~ referred to Mr Woodman's Bungalow – which you would not care for. It strikes me that young gentlemen through unsettled farming may end in selling to the Forestry – (about which neither you nor I are rabidly against) if that side of the valley, north of Brantwood, appealed to you – there might, in case of a break up, be possibility of asking about a site before a large sale – But I can quite understand your fancy for Chattenwrite, it is so

So the landscape is very old fashioned Christmas – We keep well here – I am sending you Christopher le Henry's music, it was very charming when he played it but I think too difficult for children?

I look regularly at advertisements but in vain – I think one was with reference to Mr Woodman's bungalow – which you would not care for. It strikes me that young gentleman through unsettled farming may end in selling to the Forestry – (about which neither you nor I are rabidly against) If that side of the valley, north of Brantwood, appealed to you – there might, in case of a break up, be possibility of asking about a site before a large sale. But I quite understand your fancy for Satterthaite, it is so

P.S. I did not know I had written such an immense letter (with abt 2 dy. to write) its those Pictures on the table put me out. They went to go for a walk in the sun

P.S. I didn't know I had written such an immense letter (with about 2 doz. to write) its those Pekes on the table put me out. They won't go for a walk in the snow.

Jan 28ᵗʰ 38   Castle Cottage
                      Sawry
                        Ambleside

My dear Josephine,

The music book is very lovely; it is almost pathetic to think of it in the inky hands of the printers. I do not know how such delicate drawings may be dealt with and survive. I will just tell you what happened to the other music book. I made pencil designs in pencil; and rather to my surprise Mrs Chesters said they could get pencil drawings reproduced quite well. I don't understand who did the deed but some one took a vast amount of pains to overlay my pencil with a vast crowd of little fly-like scratches in ink! which could be reproduced. The result was better than might have been expected; though they need not have taken the trouble to make a double set of foot marks where I had tried to rub out. I do not think Mrs Warne would do that. A few years ago the real way would have been for you yourself to lithograph the designs on stone — they are very like delicate lithographs. Music used to be lithographed; but it is a vanished art for commercial purposes. I believe the only plan would be for you to make stronger black outlined copies, with wash backgrounds a little stronger; and to stipulate that the resulting blocks should be printed in gray ink, to soften the effect back to the present style. They are extremely charming. The only one I do not like is Noah, where I am in doubt whether they are conventional — or live animals — or glorified toys? The coon is something quite new & effective; it arrests. I do love the little rhymes. The Robin, the Daisies, the Brave Dove are perfect for small children.

Castle Cottage
Sawrey
Ambleside

[Jan 28th 38]

My dear Josephine,

The music book is very lovely; it is almost pathetic to think of it in the inky hands of the printers. I do not know how such delicate drawings may be dealt with and survive. I will just tell you what happened with the other music book. I made preliminary designs in pencil, and rather to my surprise Messrs Chester said they could get pencil drawings reproduced quite well. I don't understand who did the deed but someone took a vast amount of pains to overlay my pencil with a vast crowd of little fly-like scratches in ink! Which could be reproduced. The result was better than might have been expected; though they need not have taken the trouble to make a double set of foot marks where I had tried to rub out. I do not think Messrs Warne would do that. A few years ago the real way would have been for you to lithograph the designs on stone – they are very like delicate lithographs. Music needs to be lithographed, but it is a vanished art for commercial purposes. I believe the only plan would be for you to make stronger black outlined copies, with washed backgrounds a little stronger; and to stipulate that the resulting blocks should be printed in gray ink, to soften the effect back to the present style. They are extremely charming. The only one I do not like is Noah, where I am in doubt whether they are conveniently live animals or glorified toys? The cover is something new and effective; it arrests. I do love the little little rhymes. The Robin, the Daisies, the Brave Dove are perfect for small children.

My own favourites are the Elphin man and the Daisies—
poor dear Pigwig! The daisies in the cradle with little
red noses are too lovely, it is a sweet drawing.
I think with one finger the tunes are much more understandable
than Mr Le Hennys — which were captivating when he
played them but more sophisticated. I do not think the tooth would
clash at all.
I think you should certainly try Messrs Warne —
Though I must confess that at present Messrs Stephens has
about as much sentiment & artistic taste as one of their own
cookery books. They have had an "inherited" stock of really good
blocks, gradually going out of copyright. It was Warne & Routledge
who published Walter Crane's "Babies Opera" years ago. Now
old fashioned & out of print; but much in advance of anything
that had come out before. I think you might safely
send Warne's this choice copy — but if it comes to hawking
round the London publishers, I should advice you to prepare
one or two sample pages with a blacker line; sad to say!
We are having very wild weather but warmer today. Snow drops
2 or 3 inches?? Mine are in flower. We enjoyed "Farmers Fells", it was
very good. Except the hymn, rather mangled; but it might be the set's
fault in transmission. I am proud to have mind Aurora. W.s were
in the cellar with a candle shelling 2 pigs eyes.
I do like the idea, it is a sweet book.
Love to you & Delmar, yrs aff.
Beatrix Heelis

My own favourites are the Elphin Man and the Daisies – poor dear Pigwig! The daisies in the cradle with little red noses are too lovely, it is a sweet drawing. I think with one finger the tunes are much more understandable than Mr Le Henrys, which were captivating when he played them but more sophisticated. I do not think the books would clash at all.

I think you should certainly try Messrs Warne – though I must confess that the present manager Mr Stephens has about as much sentiment & artistic taste as one of their own cookery books. They have had an inherited stock of really good books, gradually going out of copyright. It was Warne & Routledge who published Walter Crane's "Babus Opera" years ago. Now old-fashioned & out of print; but much in advance of anything that had come out before.

I think you might safely send Warnes this choice copy – but if it comes to hawking round the London publishers, I should advise you to prepare one or two sample pages with a blacker line; sad to say!!

We are having very wild weather but warmer today. Snow drops 2 or 3 inches?? Mine are in flower. We enjoyed "Furness Fells", it was very good. Except the hymn, rather mangled; but it might be the "Set's" fault in transcription. I am provoked to have missed aurora. W & I were in the cellar with a candle salting 2 pigs legs.

Love to you & Delmar, yrs aff

Beatrix Heelis

Castle Cottage
Sawrey

Feb 25. 35

My dear Josephine,

It was disappointing but not unexpected by me. I am afraid your originals are too delicately beautiful for the modern publishing world which caters for shop keepers at a competitive price. Goupil could have reproduced it as an art book years ago. What is to be done in face of such a world? Children deserve the best. Rather than have the vexation of seeing your work reproduced badly, I should advise you some day when you don't feel too disgusted — to re copy some of your pages in darker ink (not necessarily black, tho' black reproduces more correctly in values) and adapt them to a smaller page — or arrange to have more white upper — or, if the music publisher wants the full size minus — compress the design so that there is a wider margin. A design spread over a page the size of your sample would require a very large plate; reckoned by the square inch, its formidable. They are too good and pretty to be left in the cold of rejection. Neither do I see — though I can understand the feeling that there is something superior and more satisfying to the artist in a very choice very limited small edition — But there!! Peter Rabbit aspired to be high art, and he was passable (except the covers which I had nothing to do with and always hated), but if not high art his moderate price has at least enabled him to reach many hundreds of thousands of children, and has given them pleasure without ugliness. I do hope you will

[86]

Feb 28 38

My dear Josephine,

  I was disappointing but not unexpected by me – I am afraid your originals are too delicately beautiful for the modern publishing world which caters for shop keepers at a competitive price. Goupil could have reproduced it as an art book years ago. What is to be done in face of such a world? Children deserve the best. Rather than having the vexation of seeing your work reproduced badly, I should advise you some day when you don't feel too disgusted – to recopy some of your pages in darker ink (not necessarily black, tho' black reproduces more correct in values) and adapt them to a smaller page or arrange to have more white margins. Or, if the music publisher wants the full size music – compress the design so that there is a wider margin. A design spread over the page the size of your sample would require a very large plate; reckoned by the square inch, its formidable. They are too good and pretty to be left in the cold of rejection. Neither do I see – though I can understand – that there is something expressive and more satisfying to the artist in a very choice very limited small edition – But there!! Peter never aspired to be high art – he was passable (except the covers which I had nothing to do with and always hated), but if not high art his moderate price has at least enabled him to reach many hundreds of thousands of children, and has given them pleasure without ugliness. I do hope you will

get your book published in some form. If you recopy I
do not think you should give yourself the labour of
beautifully transcribing all the music pages. You don't know
what a dirty mess printers etc make! music on ordinary
bought music paper is good enough for them.
We have had a pleasant dry spell, with occasional still
days — Today is wild rain. I have been in Little Langdale
and Coniston on days when the fells were very lovely. On
Oxenfell on Tuesday I looked away — away — there was another
behind Bowfell, in the gap between Bowfell and Langdale Pikes.
The fells are never twice alike — There has been an
amazing show of Dick Yeadon's water colours at Kendal.
It is a tragedy that he died last summer of pneumonia
at 40 — He was 'finding himself' the last six or eight years.
He was mainly self taught, and weak in perspective which he had
obviously never learnt; but he did understand the Howgill and
Barbon & Tebay fells. Not successful in his few attempts in
the Lake district proper which one can understand as he worked
much from memory and public notes and would not remember the forms of unaccustomed
hills. When Delmar can get as much light into his studies
as poor Dick ??! I have a sketch of Skap moor before me
which reminds me of a slight sketch by David Cox which belonged
to my grandfather. All done in the spare time of Saturday Sundays
after driving a laundry van. It was a very remarkable and sad
show — closed yesterday after one week. Your husband has learnt clouds.
Light next please. He has the drawing which is the foundation.
yrs sincerely Beatrix Heelis

get your book published in some form. If you recopy I do not think you should give yourself the labour of beautifully transcribing all the music pages. You don't know what a dirty mess printers etc. make! Music on ordinary bright music paper is good enough for them.

    We have had a pleasant dry spell, with occasional still days – Today is wild rain. I have been in Little Langdale and Coniston on days when the fells were very lovely. On Oxenfell on Tuesday I looked away – away – there was another behind Bowfell, in the gap between Bowfell and Langdale Pikes. The fells were never twice alike. There has been an amazing show of Dick Yeadon's water colours at Kendal. It is a tragedy that he died last summer of pneumonia at 40. He was 'finding himself' the last six or eight years. He was mainly self taught, And weak in perspective which he had obviously never learnt, but he did understand the How Gill and Barbon & Tebay fells. Not successful in his few attempts in the Lake District proper which one can understand as he worked much from memory and pencil notes and would not remember the forms of unaccustomed hills. When Delmar can get as much light into his studies as poor Dick. ??! I have a sketch of Shap moor before me which reminds me of a slight sketch by David Cox which belonged to my grandfather. All done in the spare times of Saturday Sundays after driving a laundry van 12 hours a day during the week. It was a very remarkable and sad show – closed yesterday after one week. Your husband has learned clouds. Light next please. He has the drawing which is the foundation.

        Yrs sincerely

         Beatrix Heelis

June 26. 39   Castle Cottage
                                 Sawrey

My dear Figgins!

What an affliction – jaundice
is such a depressing illness.
You are very wise to have gone to
Hospital where you will be well
looked after, and treated, as it can
hang on for months unless thoroughly
cleared out. The last influenza
epidemic left many seedy victims; but
mainly throats. I am sorry I
cannot call to see you as Mrs Bannin
suggests, because I feel the shaking of
any long drive, although I am much

June 26 [19]39

My dear Pigwig!

What an affliction – jaundice is such a depressing illness. You are very wise to have gone to Hospital where you will be well looked after, and treated, as it can hang on for months unless thoroughly cleared out. The last influenza epidemic left many seedy victims; but mainly throats. I am sorry I cannot call to see you as Mr Banner suggests, because I feel the shaking of any long drive, although I am much

better - I went to Liverpool again this spring for a second operation, which was successful, but of course it takes time to recover after effects. We are starting hay here, but there is little to cut; its a pity fine weather should do harm. The flowers - especially climbing roses have been lovely. About "Constance Holme's" books, the scenery is sand side and the headlands between Cartmel and the Ulverston sands and estuary - she is Mrs Punchard; land-agents about Kirkby Lonsdale and Kendal. Very well written; not the cheerfulest reading for a jaundiced patient!

I hope you will soon be well again.

yours sincerely   I see from reading the letter

Beatrix Heelis   he is out of sorts too - it never rains but it pours!

better – I went to Liverpool again this spring for a second operation, which was successful, but of course it takes time to recover after effects. We are starting hay here, but there is little to cut; it's a pity fine weather should do harm. The flowers – especially climbing roses have been lovely. About "Contance Holne's" books, the scenery is Sandside and the headlands between Cartmel and the Ulverston sands and estuary – she is Mrs Punchard; land-agents about Kirkby Lonsdale and Kendal. very well written; not the cheerfulest reading for a jaundiced patient!

> I hope you will soon be well again,
>> Yours sincerely
>>> Beatrix Heelis

I see from re-reading the letter he is out of sorts too – it never rains but it pours!

Castle Cottage
Sawrey
March 7. 43   Ambleside

My dear Josephine,

Your letter of Jan 15ᵗʰ is
still reproaching me from an unanswered
bundle — they do accumulate; and
it is not always easy to find
the answer or to know how —
I think in difficult times the
true philosophy is to keep the mind
so far as possible to the trivial round
and common task, thinking as little
as may be about the things which
we cannot mend. You — not being
very strong — should read for amusement

Castle Cottage
Sawrey
Ambleside

March 7 [19]43

My dear Josephine,

Your letter of Jan 15th is still reproaching me from an
unanswered bundle – they do accumulate and it is not always
easy to find the answer or to know how – I think in difficult
times the true philosophy is to keep the mind so far as possible
to the trivial round and common task, thinking as little as may
be about the things which we cannot mend. You – not being
very strong – should read for amusement

not difficult books. The Peter Rabbit books are harmless reading for the young anyhow! not too exciting. I have been unusually well; improving by people unable to buy them, which is due to the paper quota, and most that went. I hope you and Mr Hammon have escaped flu, there has been a good deal about Sawrey & Hawkshead; not severe, but a nuisance. Mrs Armpit seriously, and the boy; but winter has thrown back work awkwardly. This spell of fine weather has been most welcome. Its just too early for snow; can I feel that tempting.

Snowdrops have been very pretty, just going over. The daffodil flytes was much in evidence last summer — and the grubs have made havoc amongst the tulips — which are seriously thinned out, half of them fallen & rotten in this garden. Its curious how grubs increase in waves. The "bracken clock" grubs are going to ruin the upland pastures if they go on spreading at their present pace. I think they had a hold in Little Langdale before they were bad here; as they are spreading, but in Coniston. We have had to plough a steep unsuitable field behind this house, because grubs + crows had skinned it bare. The turf came off in great slabs, and in spite of the crows industry the soil is full of white grubs. I don't think the weather

– not difficult books – The Peter Rabbit books are harmless reading for the young anyhow! not too exciting – I have been inundated with inquiries by people unable to buy them, which is due to the paper quota, and must wait. I hope you and Mr Banner have escaped flu, there has been a good deal about Sawrey & Hawkshead; not severe, but a nuisance – Illness amongst servants and the very wet winter has thrown back work awkwardly – This spell of fine weather has been most welcome. Its just too early for sowing corn & seeds, though tempting. Snowdrops have been very pretty, just going over. The daffodil fly was much inevidence last summer – and the grubs have made havoc amongst the bulbs – which are severely thinned out, half of them hollow and rotten in this garden – Its curious how pests increase in waves – The 'bracken clock' grubs are going to ruin the upland pastures if they go on spreading at their current pace – I think they had a hold in Little Langdale before they were bad here; and they are spreading about Coniston. We have had to plough a steep unsuitable field behind this house, because grubs & crows had skimmed it bare – The turf came off in great slabs, and in spite of the crows' industry the soil is full of white grubs. I don't think the weather

makes much difference, they are too deep in the turf to be affected by frost. You have not had much snow to look at on the fells; but sunsets have been very beautiful this winter. I wonder if the wild swans have been on the tarn. I saw a flock of tiny birds in the garden yesterday that have been strangers for several years — red polls. I also saw a pair of long tail tits. Little birds had a bad time in the big snows. We have never seen one Jenny wren since. There are twin lambs, but the main lambing time won't be till middle of April. What with shortage of petrol and flu farming is a bit of a problem! But keep smiling!

With kind regards and best wishes yrs sincerely Beatrix Heelis.

makes much difference, they are too deep in the turf to be affected by frost. You have not had much snow to look at on the fells; but sunsets have been very beautiful this winter. I wonder if the wild swans have been on the tarn. I saw a flock of tiny birds in the garden yesterday that have been strangers for several years – red polls. I also saw a pair of long tail tits – Little birds had a bad time in the big snows. We have never seen one Jenny Wren since. There are twin lambs, but the main lambing time won't be till middle of April. What with shortage of petrol and flu – farming is a bit of a problem! But keep smiling!

    With kind regards and best wishes yrs sincerely
    Beatrix Heelis

H B H —
a portrait
hows tht ??!!

mottoa
"Keep Smiling!'

───────── 13 ─────────

SOME, WHO WROTE about her didn't get within touching distance.

Just as my Aunt who adored her Cat, called it (in front of people she suspected might be thinking her sentimental) "a wicked old moggie!" and just as a very tough shepherd called some wild flowers "the little buggers" (because he could not bear to sell the plot of land where they had flourished even in his Father's days), so that if Beatrix sometimes called her husband "silly" and "old" that she meant to belittle him!

One afternoon we stayed on to see William before supper. He was late in from the office and the daylight fading. All peacefully silent except for the slight settling of the embers glowing in the fire. She moved to her old table to set it for supper. A white Linen cloth, Victorian silver and glass. In a mysterious way we became the removed watchers of a younger Beatrix lighting the Candles in readiness for the return of a personality like one of the romantic characters from the pen of Jane Austen ... and yes, when he entered the room there was an ambience about him of an eighteenth century quiet gentleman.

And so, Beloved – here is my last letter to you, written with my left, to give my right a rest for sculpture.

I've just finished a portrait of the poet John Clare for the Wordsworth Trust to look at.

The Weight of our Sins is almost ready to go the Sir Colm's International bure in Edinburgh – and lots else – but I think you will like my Solway Cross best. There will be a large one in stone to look through (and see the Lamb of God set in the centre of the First Christian Stone Circle. 'If your eye be simple, your whole body will be full of light'. But here is a little one – small enough for patients with insomnia to cuddle under the bedcloths. For the Blind the quotation is 'Blessed are they who have not seen, yet have believed'.

I think you'd smile to see me aged 98 – how I grope about, trying to pick up dropped oddments and my back bent over like a crab doing a cake walk but like you – "Keep smiling".

This film is of the little bear character invented by Josefina to amuse the children and was filmed in 1949 by the Polish film-maker Bernard Kunicki. Josefina herself is inside the bear.

14

I TOO HAVE wanted to amuse children – but the last thing I would do would be to take any animal of yours characters … so I invented a sort of Koala Bear called PATTY.

15

DEAR BEATRIX, I am adding this because I know it will please you.

My friend in London had a high top Flat. One day a pair of doves flew in. They said 'We have come to live here' (they were quite sure she would welcome them). She phoned me (somewhat frantic). I came with a box of tools and made a small Cote for them on a wide space beyond her Window sill. They were a devoted pair and had no fear of us.

They must have escaped from a Circus, because when we were sitting quiet they seemed to want to please us, and did their Tricks. The best was to perch on the Shade of a large Standing Lamp. It was quite secure but revolved at a touch. One would fly up and perch, the very charmingly and quite fast move his little feet and cause the Shade to turn at an angle – all the while looking at us for approval! They took turn and never made a mistake in a performance. There would be no question but that it was purposely done to please us.

In the end they left – but we think that when they had settled and found their way about that they went to live in one of the London Squares or Parks.

David Johns

Josefina de Vasconcellos with 'Reconciliation', Bradford University

'Reconciliation' in Japan

——— 16 ———

Beloved – many thoughts and memories of my own have been left out. One day (maybe soon) we'll catch up, or more likely, won't need to!

But there's one thing that might help some others to be rid of a trouble, if I told of it?

Last year, I came to a pass in my life when it became urgent for me to forgive any who had hurt or injured me. Luckily a very kind and understanding Pastor came to give me Holy Communion. I told him of my troubles … and that I could not manage it alone.

With the greatest help and faith, he added prayers of Release and Forgiveness to the service.

Ever since then I have felt comforted and at peace. And now there's nothing more to say except thanks – and thanks!

Here is a poem to share with you because you understand.

To One, seen but once

Still real – still dear …
Years passed,
Yet first and last,
From image clear,
in tune with Time.

Your presence in my life,
near as breath
to the Oboe,
or flowing song
to the River.

My spirit lingers
in the air of a question …
Yet still content to know,
that somewhere –
You are There.

Once upon a time

Start of *The Tale of Peter Rabbit*
by *Josefina de Vasconcellos*
(cast in J B formula)

She was loved

1
Long after,
the tongues that hissed
spite and slander,
had ceased to poison –
and Time's hour-glass,
turned by an unseen hand –
She was loved.

2
Long after
the house she knew,
had fallen –
and leaves from trees
in the Boltons
lay in golden patterns
on the pavement –
She was loved

3
Long after
the embers had died
in the grate of Hill Top parlour,
and the candles stood cold
in their sockets –
she was loved.

4
Long after,
the acorns she planted
became oak trees,
and the wild geese
returned once again
in 2003 …
She was loved …
because she greatly loved.